EASY GUIDE TO RAISING BACKYARD CHICKENS

Essential Guide to Raising Healthy, Happy
Hens - Simplified Chicken Care, Coop
Basics, and Urban Farming Techniques
for Beginner Homesteaders

DALE G. FAUSTINO

CHAPTER ONE

INTRODUCTION

Raising backyard chickens has become highly popular pastime for families and individuals alike. Not merely do chickens supply a renewable supply of eggs that are fresh, though they also provide companionship and a unique link with the natural world. Nevertheless, beginning with backyard chickens needs careful planning and consideration to guarantee the health and also well being of the feathered friends of yours and compliance with local laws. This comprehensive guide is going to walk you through the important steps to get going with backyard chickens.

Selecting Suitable Chicken Breeds

Egg-laying Breeds

One of the primary decisions you will have making when starting off with backyard chickens is selecting the best chicken breeds for the needs of yours. Egg-laying breeds are particularly bred for their capability to create a consistent source of eggs.

Some common egg laying breeds include:

1. Rhode Island Red

The Rhode Island Red is noted for its superb egg laying features, with hens producing brown eggs. They're hardy birds and make a wonderful option for beginners.

2. Leghorn

Leghorns are prolific levels of cream eggs and are recognized for their alert and active nature. They're perfect in case you are searching for high egg production.

3. Sussex

Sussex chickens lay brown eggs and therefore are recognized for their friendly disposition. They're a great choice for families with kids.

4. Plymouth Rock (Barred Rock)

Plymouth Rocks are dual purpose birds but are also great egg layers. They're well-suited and docile for backyard environments.

Dual-purpose Breeds

In case you are considering not just eggs but additionally meat production, dual purpose breeds are a great alternative. These breeds are flexible and will provide both meat and eggs. Some common dual purpose breeds include:

1. Orpington

Orpington chickens are big, welcoming birds recognized for their great egg production and quality meat. They are available in colors that are various, including black and buff.

2. Australorp

Australorps are recognized for keeping the world record for egg production - 364 eggs in 365 days. They're also good meat birds and also have a relatively calm temperament.

3. Wyandotte

Wyandottes are dual purpose birds recognized for their attractive feather patterns egg laying abilities. They create brown eggs and also have a durable build.

Local Regulations and Zoning Before you get started on a backyard chicken adventure, it is essential to investigate and also understand the local regulations of yours and zoning regulations concerning poultry keeping. These laws are able to differ widely from one spot to another and could cover aspects such as:

1. Permits and Licensing

Some areas require licenses or permits to keep chickens. These documents frequently specify the amount of chickens allowed along with other related regulations.

2. Setback Requirements

Setback requirements determine the distance your chicken coop should be from property lines, neighboring homes, and public roads. These rules wish to reduce disturbances to neighbors.

3. Noise Ordinances

Chickens are generally noisy, particularly when they are laying eggs or even disturbed. Local noise ordinances could limit the time during which chickens are able to make noise.

4. Safety and Health Regulations

Local authorities contains safety and health laws set up to stop the spread of diseases among chickens and ensure appropriate waste disposal.

5. Nuisance Laws

Nuisance laws address problems like smells, flies, and insects that will end up from keeping chickens. Complying with these laws can help maintain relations that are good with neighbors.

To discover about local regulations, contact the city of yours or maybe county government office, and go to the website of theirs. You might also need to talk to friends to make sure they are at ease with your choice to raise chickens.

Setting up Your Chicken Coop

After you have selected your chicken breeds and also checked local regulations, it is time to prepare your chicken coop. The coop is when the chickens of yours are going to sleep, lay eggs, and find protection out of the elements. Good planning is crucial to create a comfortable and safe environment for the birds of yours.

Coop Design and Size

The dimensions of your coop depends upon the quantity of

chickens you intend to keep. A basic guideline is providing a minimum of 2 3 square feet of interior space per chicken and 8 10 square feet of outdoor space per chicken. Nevertheless, more room is always better.

2. Design

Think about the following design elements when considering your coop:

A. Ventilation

Proper ventilation is essential to avoid moisture buildup and respiratory problems in chickens. Include vents near the top to permit superior airflow while defending against drafts.

B. Nesting Boxes

Provide nesting boxes for your hens to lay eggs. Each package must be about 12x12 inches and positioned from the earth to discourage broodiness.

C. Roosting Bars

Chickens love to perch at night. Install roosting bars at a level of aproximatelly 2 4 feet above the coop flooring, with sufficient room for every bird.

D. Easy Cleaning

Plan for cleaning that is easy by incorporating features like removable droppings trays or maybe permission to access the coop flooring.

E. Predator Protection

Ensure your coop is predator proof by making use of strong materials, burying wire mesh to prevent digging animals, and locking coop doors properly when it's dark.

Location and also Placement

Place the coop of yours exactly where it is able to get ample sunshine throughout the day. Sunlight not just offers heat but also helps reduce water and helps to keep the coop dry.

2. Accessibility

Make certain your coop is readily accessible for everyday chores like feeding, egg collection, and watering. Convenient access is going to save you effort and time.

3. Proximity to the House

Think about locating your coop reasonably near the house of yours to allow it to be easier to evaluate the chickens of yours and also supply them with care.

4. Consider The Neighbors of yours

Be mindful of the neighbors of yours when choosing the coop location. Make sure it does not block the views of theirs or create disturbances.

5. Drainage and Terrain

Select a website with good drainage to stop flooding during heavy rains. Elevated areas with well draining soil are perfect.

To conclude, starting out with backyard chickens could be a rewarding experience, though it takes thorough consideration and planning of elements as chicken breeds, local laws, coop size and style, and placement. By taking time to make educated choices and also develop a comfy setting for the feathered friends of yours, you are able to appreciate the advantages of the joy and fresh eggs of raising chickens in the own backyard of yours while becoming a conscientious neighbor and complying with local laws.

CHAPTER TWO

CARING FOR YOUR BACKYARD CHICKENS

Raising backyard chickens could be a satisfying and rewarding experience, though it in addition comes with responsibilities. Appropriate attention and care are vital to make sure your chickens are productive, happy, and healthy. With this extensive manual, we will delve into different issues with taking care of the backyard chickens of yours, covering topics like feeding, supplying water that is clean, monitoring chicken health, handling and interacting with chickens, and understanding the behavior of theirs.

Providing for The Chickens of yours

Feeding the chickens of yours a nutritious and balanced diet plan is fundamental to their productivity and health. Let us explore the crucial aspects of feeding the feathered friends of yours.

Kinds of Chicken Feed one. Starter Feed

Starter feed is especially developed for young chicks, usually from day one to eight weeks old. It has higher protein levels (18 20 %) to allow for fast development and feather development.

2. Grower Feed

Grower feed is designed for chickens aged eight to eighteen weeks. It's somewhat reduced protein content (around 16 18 %) than starter feed and also offers the required nutrients for good development.

3. Layer Feed

Layer feed is created for adult hens which are definitely laying eggs. It has calcium (usually around 16 18 %) to allow for good eggshell production. Layer feed must be unveiled when your hens begin laying, usually around 18 20 weeks of age.

4. All Purpose Feed

Some chicken keepers prefer utilizing an all purpose feed which may be given to chickens of ages. This feed generally has a moderate protein level (around 16 18 %) and also may simplify feeding logistics.

Nourishing Schedule

Establishing a regular feeding routine is essential for your chickens' well being. Here is a fundamental feeding schedule to consider:

a. Morning

Offer water that is fresh and check on the chickens of yours. Offer a percentage of their everyday feed to make sure they begin the day with power.

b. Midday

Chickens may perhaps forage and peck for vegetation and insects during the morning. If they've a chance to access a free range area, they'll usually find the own food of theirs.

c. Evening

Before dusk, provide another feeding to make sure they've food that is enough to last throughout the night. Chickens do not consume in the dark, therefore it is crucial to offer their evening meal while there is always daylight.

Treats and Supplements

While commercial chicken feed provides nutrients that are essential, you are able to additionally augment your chickens' diet plan with treats as well as dietary supplements to add variety and also address specific needs:

1. Vegetables and Fruits

Chickens like a variety of fresh fruits and vegetables, like lettuce, watermelon, cucumbers, and apples. These treats provide minerals and vitamins.

2. Seeds and grains

Scratch grains, cracked corn, as well sunflower seed are presented as rare treats. These treats are energy rich and also could be spread to encourage natural foraging behavior.

3. Grit

Grit is compact, hard particles as rocks or maybe shells that chickens have to help digestion. It will help them grind down the food of theirs in their gizzard.

4. Calcium

Crushed eggshells or oyster shell could be offered to make certain hens have sufficient calcium for strong eggshells, particularly when feeding layer feed.

5. Insects and mealworms

Mealworms and insects are fantastic sources of protein for chickens. They may be presented as treats and sprinkled in their coop to inspire foraging.

Note: Treats as well as supplements must be given in small amounts, as increased treats could result in imbalances in the diet of theirs. Stick with a basic rule of offering treats as no greater than ten % of the daily food intake of theirs.

Providing Water that is clean

Access to clean, water that is fresh is crucial for your chickens' overall well-being and health. Here is how you can make sure they've the water they need:

1. Clean and Refill Daily

Look at the water containers every day, clean them as necessary, and refill with water that is fresh. Chickens are generally messy, and the water of theirs could easily get contaminated with droppings and dirt.

2. Prevent Freezing

In cool weather, provide heated waterers or maybe check water sources often to make sure they do not freeze in cold weather. Dehydration could be a major problem in weather that is cold.

3. Keep Waterers Off the Ground

Elevating waterers are able to avoid contamination and maintain the water cleaner. Additionally, it cuts down on the chance of spills.

4. Multiple Water Sources

If you've a sizable flock, think about providing several water sources to stay away from crowding and make certain all chickens are able to access water easily.

Monitoring Chicken Health

Frequent monitoring of your chickens' overall health is vital for early detection of any problems. Here is how you can monitor their well-being:

Symptoms of Illness one. Change in Behavior
Observe your chickens' behavior every day. Any sudden changes, like unusual aggression, isolation, or lethargy, might indicate illness.
2. Abnormal Droppings

Examine the droppings of theirs for symptoms of diarrhea, blood, or perhaps abnormal colors. Healthy droppings must be brown and firm.

3. Respiratory Issues

Listen for coughing, or wheezing, sneezing, as these could indicate respiratory infections.

4. Changes in Eating or even Drinking Habits

Chickens that stop eating or even drinking could be unwell. Monitor their water and food usage.

5. Feather Issues

Check for feather loss, feather pecking, or even symptoms of parasites on the skin of theirs.

Preventive Care

1. Quarantine New Birds

When introducing brand new chickens to the flock of yours, quarantine them for no less than 2 days. This can help stop the spread of diseases.

2. Vaccinations

Consult with a poultry veterinarian to decide whether vaccinations are needed for typical chicken diseases in the area of yours.

3. Parasite Control

Implement a standard parasite management program, including treatments for external and internal parasites as mites and worms.

4. Clean Coop

Maintain a thoroughly clean coop to stop the buildup of waste and also the attraction of disease and pests.

Socializing and handling with Chickens

Chickens are generally friendly and like human interaction, though it is crucial to handle them appropriately: and gently

1. Approach Calmly

When approaching the chickens of yours, do so steadily and stay away from abrupt movements. Let them become used to the presence of yours.

2. Hold Properly

When grabbing a chicken, use both hands to help its entire body. Never grab or squeeze them forcefully.

3. Spend time with Them

Spend time together with your chickens to socialize them. Hand-feeding treats can help develop trust and make a positive connection with the presence of yours.

4. Stay away from Chasing

Chickens may become stressed if chased or maybe pursued. Should you have to capture one, do so steadily and carefully.

Understanding Chicken Behavior

Understanding chicken behavior is crucial for their well-being and the ability of yours to take care of them effectively:

1. Pecking Order

Chickens have a cultural hierarchy referred to as the pecking order. It is crucial to understand and admire this hierarchy within the flock of yours.

2. Dust Bathing

Chickens take dust baths to thoroughly clean their skin and feathers. Supply them with a designated dust bath area, like a package of dried out dirt or sand.

3. Roosting

Chickens love to roost at night. Ensure your coop has adequate roosting bars at a suitable level.

4. Egg Laying Behavior

Hens usually display particular behaviors before laying eggs, like nesting in exactly the same area. Provide ideal nesting boxes for them. To conclude, taking care of backyard chickens involves different aspects, which includes feeding, providing water that is clean, monitoring understanding, socializing, handling, and health the behavior of theirs. By adhering to these recommendations and paying attention for your chickens' behaviors and needs, you are able to make sure a healthy and happy flock which provides you with eggs that are fresh and companionship for decades to come.

Keep in mind that every chicken has its own preferences and personality, therefore getting to find out your birds individually will enhance the ability of yours to take care of them properly.

CHAPTER THREE

CHICKEN HOUSING AND ENVIRONMENT

Raising chickens in the backyard of yours could be an enjoyable and rewarding experience, though it takes good housing and a suitable setting to guarantee the overall health, safety, and well being of the feathered friends of yours. With this extensive manual, we will check out the crucial areas of chicken housing and environment, such as coop upkeep and cleaning, nesting boxes and roosting bars, good insulation and ventilation, predator protection, and outside runs and free range options.

Coop Maintenance and Cleaning

A well-maintained and clean chicken coop is crucial for the and happiness of the chickens of yours. Regular cleaning and maintenance responsibilities assist in preventing disease, minimize odors, and also ensure a comfy living room.

1. Daily Tasks

a. Removing Droppings

Each and every day, take out some wet or soiled droppings and bedding from the coop flooring and nesting boxes. This prevents the buildup of ammonia and minimizes the chance of illness.

Water and b. Checking Food
Check out and also refill food and water containers daily to make sure your chickens obtain new supplies.

2. Weekly Tasks

a. Changing Bedding
When a week, replace all of the bedding inside the coop to maintain freshness and cleanliness. Common bedding options are straw, pine shavings, and hay.

b. Cleaning Nesting Boxes
Inspect and clean the nesting boxes on a regular basis, removing any

soiled or perhaps broken eggs. Add new bedding material to hold the boxes inviting for egg laying.

3. Monthly Tasks

a. Deep Cleaning

Conduct a deep cleaning of the whole coop and nesting boxes the moment a month. This involves scrubbing checking, disinfecting, and surfaces for signs of damage or wear.

b. Inspecting Hardware

Regularly inspect hardware, like latches, locks, and also hinges, to make sure they're protected and functioning properly. This aids in preventing predator access.

Nesting Boxes and Roosting Bars

Nesting boxes and roosting bars are

crucial ingredients of your respective chicken coop that offer comfort, security, and also a spot for your hens to lay eggs and roost.

1. Nesting Boxes

a. Number of Boxes

Provide one nesting box for every 4 to 5 hens. Having plenty of boxes reduces competition and also motivates hens to lay eggs within the designated location.

b. Bedding

Line the nesting boxes with fresh bedding material, like straw or even pine shavings. This causes a comfortable and inviting room for egg laying.

c. Location

Position nesting boxes in a peaceful and dimly lit part of the coop to offer hens a sense of security and privacy. Elevated nesting boxes are able to deter egg eating and also provide cleaner eggs.

2. Roosting Bars

a. Placement

Install roosting bars within the coop at a level of 2-4 feet above the soil. Chickens prefer roosting from the ground to feel really protected from potential predators.

b. Spacing

Make sure there's ample room on the roosting bars for all the chickens of yours to comfortably roost with no overcrowding.

c. Cleaning

Clean roosting bars frequently to take out droppings and also stop the buildup of ammonia smells.

Adequate Insulation and Ventilation Ventilation and insulation play important roles in keeping a healthy and comfortable environment inside the chicken coop, particularly in different environmental conditions.

1. Ventilation

a. Importance

Proper ventilation helps control temperature, eliminate excess moisture, and keep quality of the air inside the coop. Additionally, it cuts down on the chance of respiratory problems in chickens.

b. Placement

Install vents near the top of the coop to enable warm, moist air to escape. Make sure there's sufficient airflow without developing drafts.

2. Insulation

a. Cold Climates

In colder regions, enveloping the coop is able to keep chickens warm. Use insulating materials as fiberglass or maybe foam board over the walls and ceiling.

b. Hot Climates

In warm climates, concentrate on providing shade and fresh airflow rather compared to insulation. Adequate shade and ventilation is able to avoid heat stress in chickens.

Predator Protection

Protecting the chickens of yours from predators is a high priority. Predators are able to consist of raccoons, foxes, coyotes, birds of prey, as well as neighborhood dogs. Here is how you can maintain your flock safe:

1. Secure Locks

Pick locks and latches on nesting boxes and coop doors to prevent access that is easy for predators. Make sure that doors are properly closed at night.

2. Reinforced Walls

Make use of sturdy materials for coop wall space and floors to stop digging animals from burrowing into the coop. Bury wire mesh within the perimeter to deter digging.

3. Elevated Coop

Elevating the coop from the earth is able to help make it much more complicated for predators to get access. Ensure there aren't any openings or gaps underneath the coop.

4. Nighttime Security

Install motion activated lights or perhaps alarms to prevent nocturnal predators. You are able to also think about locking your chickens inside a safe run or coop at night.

5. Lock Up Food

Store chicken feed properly in rodent proof containers to avoid attracting insects which may, in turn, attract predators.

Outside Runs and also Free Range Options Providing the chickens of yours with permission to access the exterior not just improves the lives of theirs but also allows them to forage for healthy foods. Here are choices for outdoor runs and free range environments:

1. Outdoor Runs
a. Enclosed Runs
Create enclosed outdoor runs with secure fencing to safeguard chickens from predators. Ensure there's sufficient shade, shelter, and water.

b. Grazing Area

Include a grassy area within the run for chickens to graze. Rotate the place of the run occasionally to keep overgrazing.

c. Dust Bath Area
Provide a designated location with loose sand or soil for chickens to shoot dust baths. This can help make their feathers clean and wholesome.

2. Free Range Options
a. Supervised Free Range
Allowing chickens to free range under supervision in the backyard of yours could be a gratifying experience. Ensure your property is

protected, and monitor for possible dangers.

b. Electric Fencing

Use portable electric fencing to produce a protected free range location which may be moved around the property of yours.

c. Tethering

Tethering chickens working with a chicken harness or maybe mobile coop is able to offer a controlled free range experience while protecting them from predators.

To conclude, providing suitable housing and keeping a proper environment for the backyard chickens of yours is crucial for their well being. Coop maintenance, predator protection, insulation, proper ventilation, roosting bars, nesting boxes, and outside runs or maybe free range options all promote the all around health and happiness of the flock of yours. By adhering to these recommendations and consistently inspecting and looking after your chicken housing and environment, you are able to build a comfortable and safe home for the chickens of yours, making sure they lead productive and happy life in the backyard of yours.

CHAPTER FOUR

MAXIMIZING EGG PRODUCTION AND RAISING CHICKS

Raising backyard chickens not just gives a supply of eggs that are fresh but also provides the chance to hatch and raise adorable baby chicks. With this extensive manual, we are going to explore different areas of egg production and chick rearing, including maximizing egg production, breeding raising, hatching, incubation, and chickens chicks.

Maximizing Egg Production

Achieving an abundant and consistent supply of eggs from the backyard chickens of yours requires attention to a number of factors. Here is how you can optimize egg production:

1. Artificial Lighting

Chickens depend on the length of daylight to cause egg laying. As daylight decreases in the autumn and winter, egg production is likely to drop. To fight this, you are able to use artificial lighting to extend the time of light in the coop. Here is how:

a. Light Source

Use a brilliant light source in the coop, like LED light bulbs. Use timers to offer 14 16 hours of light each day, mimicking extended summer days.

b. Consistency

Keep a regular lighting schedule, with the gentle turning on earlier in the early morning and flipping off in the evening.

c. Gradual Changes

When setting the lighting time, make changes slowly to stay away from stressing the chickens.

2. Nesting Materials

Offering comfy and also attractive nesting boxes is crucial for helping hens to lay eggs in a designated location. Here is what you are able to do:

a. Bedding

Line nesting boxes with soft and clean bedding material, like straw, hay, or perhaps pine shavings. This causes a comfortable space for egg laying.

b. Privacy

Ensure nesting boxes are placed in a peaceful and dimly lit part of the coop. Hens favor privacy when laying eggs.

c. Regular Cleaning

Clean nesting boxes on a regular basis, removing soiled bedding and updating it with new material. This encourages hens to keep utilizing the boxes.

3. Storing and collecting Eggs

Collecting eggs promptly and keeping them properly is essential to keep their quality: and freshness

a. Frequency

Collect eggs at least twice or once one day to stop them from getting dirty or even cracked.

b. Clean Eggs

If an egg is soiled, lightly clean it using a dry cloth. Stay away from washing eggs with water, because it is able to eliminate the defensive bloom.

c. Storage

Store eggs in a dry and cool place, preferably at a temperature of around 45°F (7°C). Use a separate egg carton or maybe box to stop them from absorbing strong smells.

d. Labeling

Consider labeling eggs with the day they had been collected to keep monitor of freshness.

Breeding Chickens

Breeding chickens involves the procedure of mating roosters and hens to create fertilized eggs, which may be incubated to hatch brand new chicks. Here is a description of the breeding process:

1. Natural Mating
a. Hen-to-Rooster Ratio
Maintain a suitable ratio of hens to roosters to ensure effective mating. One rooster can typically manage 8 10 hens.

b. Observation
View the actions of the flock to guarantee that mating is occurring. Roosters usually conduct a mating dance and might be seen mounting hens.

c. Fertilization

Fertilized eggs have a white-colored area referred to as a blastoderm over the egg yolk. These eggs could be utilized for incubation.

2. Incubation and Hatching

Incubation is the procedure of artificially simulating the conditions needed for eggs to hatch into chicks. Here is just how it works:

a. Incubator

Buy a good quality incubator with precise temperature and also humidity controls. Stick to the manufacturer's directions for setup.

b. Egg Selection

Select fertile eggs for incubation. You are able to look for fertility by utilizing a procedure called "candling," that entails shining a brilliant light through the egg to find out if the blastoderm is present.

c. Humidity and Temperature

Maintain constant temperature and moisture levels in the incubator, following the standards for the particular chicken breed you're hatching.

d. Turning Eggs

Regularly flip the eggs in the incubator to stop the embryo from following the layer membrane. Many contemporary incubators have automatic turning mechanisms.

e. Hatching

Chicks typically hatch after twenty one many days of incubation. Offer a specific room of the hatching chicks, known as a "hatcher," with somewhat larger humidity.

f. Care for Newborn Chicks

After the chicks hatch, move them to some brooder setup for appropriate care and feeding.

Increasing Chicks

Raising baby chicks is a rewarding and delightful experience. In order to ensure their well-being and health, follow these guidelines:

1. Brooding Setup

a. Brooder Box

Prepare a brooder box or even location in a safe, draft-free location. Use a heat source, like a heating lamp or maybe heating plate, to keep a temperature of around 95°F (35°C) just for the very first week. Steadily lower the heat by 5°F (2.8°C) every week until they're completely feathered.

b. Bedding

Line the bottom part of the brooder with fresh bedding material like pine straw or shavings. Stay away from using paper, as it may become slippery.

c. Water and Food

Provide chicks with a chance to access water that is fresh and chick starter feed. Use shallow containers to keep drowning.

2. Chick Care and Feeding

a. Handling

Handle chicks carefully and stay away from excessive stress. Frequent handling will help them start to be much more comfortable around humans.

b. Feeding

Chicks should be fed a healthy chick starter feed that offers the required nutrients for growth. Ensure they often have access to fresh water and feed.

c. Supplements

Chicks might gain from vitamin and electrolyte supplements during the early days of theirs. Consult with a poultry veterinarian for direction.

d. Observation

Regularly see chicks for symptoms of illness, injury, and any problems with the feed of theirs and water resources. Early detection is vital for their well being.

e. Gradual Transition

As chicks grow and produce feathers, they are able to be slowly introduced to outside environments. Make sure they've shelter, protection from predators, and also permission to access a protected outside run or maybe free range area.

To conclude, raising backyard chickens offers the chance not just to have eggs that are fresh but additionally to see the pleasure of hatching and also raising baby chicks. Maximizing egg production calls for careful control of lighting, egg collection, and nesting materials. Breeding chickens calls for keeping appropriate proportions, observing mating behavior, and also controlling the incubation and also hatching process. Raising chicks involves establishing a secure and comfy brooding atmosphere, supplying appropriate care and feeding, and steadily transitioning them to outside spaces as they develop. By adhering to these guidelines, you are able to build healthy flock and a thriving that brings adorable chicks

and both eggs into the backyard of
yours.

www.ingramcontent.com/pod-product-compliance
Lightning Source LLC
Chambersburg PA
CBHW062359290526
45794CB00003B/1015